EXPLORING OUR OCEANS

INDIAN OCEAN

by Avery Toolen

pogo

Ideas for Parents and Teachers

Pogo Books let children practice reading informational text while introducing them to nonfiction features such as headings, labels, sidebars, maps, and diagrams, as well as a table of contents, glossary, and index.

Carefully leveled text with a strong photo match offers early fluent readers the support they need to succeed.

Before Reading

- "Walk" through the book and point out the various nonfiction features. Ask the student what purpose each feature serves.
- Look at the glossary together. Read and discuss the words.

Read the Book

- Have the child read the book independently.
- Invite him or her to list questions that arise from reading.

After Reading

- Discuss the child's questions. Talk about how he or she might find answers to those questions.
- Prompt the child to think more. Ask: What are some of the Indian Ocean's unique habitats? What makes them unique?

Pogo Books are published by Jump!
5357 Penn Avenue South
Minneapolis, MN 55419
www.jumplibrary.com

Library of Congress Cataloging-in-Publication Data

Names: Toolen, Avery, author.
Title: Indian Ocean / by Avery Toolen.
Description: Minneapolis, MN: Jump!, Inc., [2023]
Series: Exploring our oceans | Includes index.
Audience: Ages 7-10
Identifiers: LCCN 2021051658 (print)
LCCN 2021051659 (ebook)
ISBN 9781636907741 (hardcover)
ISBN 9781636907758 (paperback)
ISBN 9781636907765 (ebook)
Subjects: LCSH: Indian Ocean—Juvenile literature.
Classification: LCC GC721 .T66 2023 (print)
LCC GC721 (ebook)
DDC 551.46/15—dc23/eng/20211204
LC record available at
https://lccn.loc.gov/2021051658
LC ebook record available at
https://lccn.loc.gov/2021051659

Editor: Jenna Gleisner
Designer: Emma Bersie

Photo Credits: lemga/iStock, cover (manatee); acro_phuket/Shutterstock, cover (background); Richard Witcombe/Shutterstock, 1; Shutterstock, 3; KKKvintage/Shutterstock, 4; JonMilnes/Shutterstock, 5; World History Archive/Alamy, 6-7; Peter Hofstetter/Alamy, 8-9; Paul Mayall Australia/Alamy, 10-11tl; elleon/Shutterstock, 10-11tr; wilar/Shutterstock, 10-11bl; Shivang Mehta/Alamy, 10-11br; V_E/Shutterstock, 12; COULANGES/Shutterstock, 13; panoglobe/Shutterstock, 14-15; arabianEye FZ LLC/Alamy, 16-17; Reinhard Dirscherl/Alamy, 18; Plamen Galabov/Shutterstock, 19; Suzanne Long/Alamy, 20-21; Rich Carey/Shutterstock, 23.

Printed in the United States of America at Corporate Graphics in North Mankato, Minnesota.

TABLE OF CONTENTS

CHAPTER 1

WARM WATERS

Which is the warmest of Earth's five oceans? It is the Indian Ocean! Its temperature can reach 82 degrees Fahrenheit (28 degrees Celsius). This makes it a great ocean for divers!

The Red Sea is part of this ocean. It is home to the Blue Hole. Divers explore this area. It is more than 328 feet (100 meters) deep!

The Indian Ocean is made up of salt water. It is surrounded by several **continents**. They are Asia, Africa, and Australia. From Africa, fresh water flows in from the Zambezi River **Delta**.

TAKE A LOOK!

Where is the Indian Ocean? Take a look!

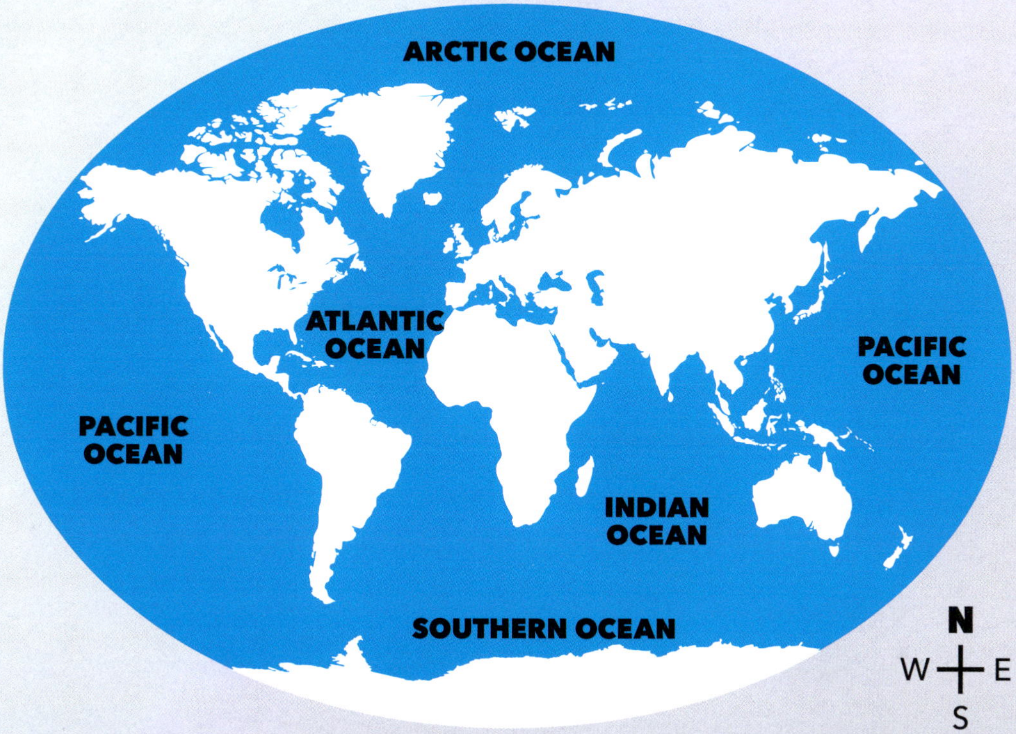

Storms are common. **Tropical cyclones** form above warm water. They cause a lot of damage when they reach land.

Underwater **earthquakes** cause tsunamis. These big waves also cause damage.

DID YOU KNOW?

Climate change is warming Earth's oceans. The Indian Ocean is warming faster than any other. Storms are becoming more frequent.

Unique **landforms** are along this ocean's coasts. Cliffs, **dunes**, **barrier islands**, and **estuaries** are some. They help protect the coasts from flooding. They provide **habitats** for wildlife, too.

cliffs

dunes

barrier islands

estuary

CHAPTER 2

ANIMALS AND PLANTS

Shallow parts of this ocean are home to **coral reefs**. Sea stars, crabs, fish, and more live here.

Madagascar is an island in this ocean. Many plants and animals that live here are found nowhere else! The ring-tailed lemur is one.

The Sundarbans is a **mangrove** forest. It is on the coast of Bangladesh and India.

Mangrove tree roots catch rocks, sand, and dirt. This helps prevent **erosion**. The roots also catch **pollutants**. This keeps them from reaching the ocean.

DID YOU KNOW?

Mangrove forests are home to many animals. Birds eat fish and crabs. Bengal tigers hunt wild boar. Sea turtles and crocodiles swim the waters. Fish swim among the tree roots.

roots

dugong

Seagrass beds are in shallow water near the shores. Dugongs swim here. The grasses provide shelter for young fish and sea turtles.

DID YOU KNOW?

Seagrasses need clear water to live. Why? This lets sunshine reach them.

CHAPTER 3

EXPLORING THE SEAS

People may have explored the Indian Ocean more than 4,000 years ago! Today, people scuba dive. They visit sunken ships.

The Persian Gulf is near the Middle East. This area of the Indian Ocean produces a lot of oil. Oil rigs drill for it. Why? It is under the ocean floor. Sometimes oil spills happen. These harm wildlife.

research boat

Scientists also explore here. They study coral reefs. They study the deep sea, too. How? They use remotely operated vehicles (ROVs). These reach areas that are difficult for humans to get to.

The Indian Ocean is full of life. What would you like to explore?

DID YOU KNOW?

The Indian Ocean's deepest point is in the Sunda **Trench**. The ocean floor is more than 4.6 miles (7.4 kilometers) deep here!

ACTIVITIES & TOOLS

TRY THIS!

MAKE A HABITAT DIORAMA

The Indian Ocean is home to many unique coastal habitats. Learn more about them with this fun activity!

What You Need:

- computer and books for researching
- notebook
- cardboard shoebox
- modeling clay
- string or yarn
- scissors
- tape

1. **Pick an Indian Ocean habitat you would like to learn more about. Research it online or use books to learn more. Find out what animals and plants live there. Learn about any unique geologic features. Write down your notes.**
2. **Find a cardboard shoebox. Remove the top. Lay your shoebox on its side lengthwise. This will be your diorama base.**
3. **Use modeling clay to sculpt different parts of the habitat, including animals, plants, and geologic features.**
4. **Place them in the shoebox. You can use tape and string or yarn to hang them from the top. This can make your fish and other ocean wildlife look like they're swimming!**
5. **Share your diorama with a friend or classmate. What can you teach them about this Indian Ocean habitat?**

GLOSSARY

barrier islands: Long, sandy islands near shore.

climate change: Changes in Earth's weather and climate over time.

continents: The seven large landmasses on Earth.

coral reefs: Long lines of coral that lie in warm, shallow waters.

delta: An area of land shaped like a triangle where a river enters the sea.

dunes: Sand hills formed by wind or tides.

earthquakes: Sudden, violent shakings of the earth that may cause damage.

erosion: The wearing away of something by water or wind.

estuaries: The wide parts of rivers, where they join the ocean.

habitats: Places where certain animals or plants are usually found.

landforms: Natural features of land surfaces.

mangrove: A kind of tree that grows in dense forests along estuaries, in salt marshes, and along coasts.

pollutants: Substances that contaminate other substances.

shallow: Not deep.

tropical cyclones: Storms that form above warm waters and have very strong winds that rotate and cause damage.

trench: A long, narrow hole in the ocean floor.

INDEX

TO LEARN MORE

Finding more information is as easy as 1, 2, 3.

1. Go to www.factsurfer.com
2. Enter "IndianOcean" into the search box.
3. Choose your book to see a list of websites.

FACT SURFER